CONSOLIDATED

B-32 DOMINATOR

HEAVY BOMBER

B-32 DOMINATOR, SECOND STRING TO THE BOEING B-29 SUPERFORTRESS

INTRODUCTION

Intended as a long-range strategic bomber, the B-32 was, in fact, a fail-safe in case the development of the revolutionary Boeing B-29 Superfortress was to run into insolvable problems.

Despite many teething problems, the B-29 finally went into production with the B-32 no longer needed — or so it seemed. However, although the B-32 also had its complications, production was started before the end of the Second World War, and it was even used operationally during the final months of the war against Japan.

The B-32 Dominator, second string to the Boeing B-29 Superfortress.

BOMBER, LONG RANGE

As early as the 1930s, the USAAC was already interested in a heavy bomber capable of intercontinental missions, with a range of 5,000 miles. The specification "Project A" for such a bomber was drafted in mid-1933. Boeing and Martin responded with designs, but only Boeing was given the go-ahead to build its design, initially known as Experimental Bomber Long Range (BLR). The resulting XB-15 (model 294) featured many revolutionary systems, but it was judged to be underpowered and too slow during testing. In the meantime, Boeing had been working on model 299 (which is widely known as B-17 Flying Fortress). This type featured a significantly more modest range but was a much more viable combat aircraft.
By mid-1935, USAAC had reformulated its specifications into "Project D", which resulted in designs by Douglas and Sikorsky. The former was contracted to build its design, the XB-19, but rapid aeronautical developments rendered it obsolete before its first flight. When XB-19 first flew, Europe had already been engulfed by war. The XB-15 and XB-19 remained prototypes since their performances were disappointing. Boeing's B-17 and Consolidated's B-24 were taken into production, but neither were suitable for very long-range missions.

The Boeing XB-15: underpowered and quickly overtaken by military thinking.

Douglas XB-19, displaying its huge 212-feet wingspan. It found a job as an engine test bed and was converted into a transport aircraft before being finally scrapped in 1946.

CONSOLIDATED AIRCRAFT CORPORATION – AN OVERVIEW

Consolidated Aircraft was formed on 29 May 1923 by Reuben Fleet, when he took over Gallaudet Aircraft and acquired the production rights of the Dayton-Wright Company. The company was further expanded by acquiring Thomas-Morse (1928) and Fleet Aircraft (1929). The company was located in Buffalo, New York, but relocated to San Diego, California, in 1935. On 17 March 1943, the company merged with Vultee Aircraft, forming the Consolidated Vultee Aircraft Corporation.

Model 1, of which 221 were built in different variants.

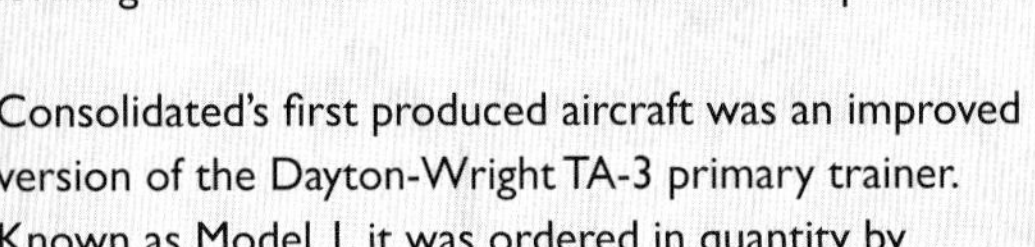

Consolidated's first produced aircraft was an improved version of the Dayton-Wright TA-3 primary trainer. Known as Model 1, it was ordered in quantity by USAAC as PT-1 and was powered by a Wright-Hispano engine. Model 2 started out as a navalised variant of PT-1, known as NY-1. It was produced in several variants with continuous improvements and differing radial engines. A final development would enter Army Air Corps service as PT-3 (Model 12) and as O-17 (company designation Model 7) with the United States National Guard. It was exported in modest numbers to several South American countries.

The XPY-1 design was built in series by the Glenn L. Martin company as Martin P3M-1 and P3M-2.

Left: P2Y-2 series aircraft featured wing-mounted engine nacelles and NACA engine cowls.

Model 9 was the first in a line of successful flying boat designs. The parasol-winged aircraft was designed in 1928 and featured an aluminium hull. The type was designed for the US Navy and early production contracts were snatched away by the Glenn Martin company, which built 10 examples. After losing the Navy contract,

Consolidated created a commercial variant, Model 16, named Commodore. It had more powerful engines, a redesigned hull shape and an enclosed cockpit for the crew. It could carry a maximum of 32 passengers. Pan Am operated the Commodore on its New York-Buenos Aires line. The Commodore, in turn, was further developed into the (X)P2Y patrol aircraft. It retained the enclosed cockpit but became a sesquiplane, by the addition of a stubby lower wing. A total of 78 were built, operating with the US Navy up to December 1941.

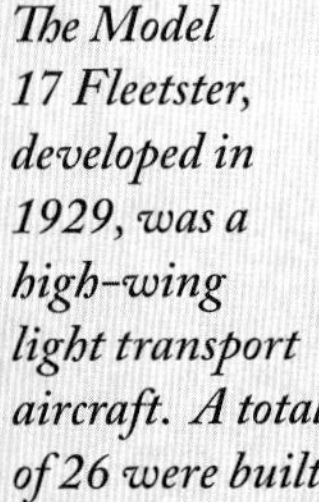

The Model 17 Fleetster, developed in 1929, was a high-wing light transport aircraft. A total of 26 were built, including a single navalised dive bomber (Model 18, shown here) and three examples of a parasol-winged variant (Model 20).

Model 21 was an aerodynamically refined version of the PT-3. A score of sub-variants was built, totalling 41 aircraft.

The design of the promising Detroit-Lockheed XP-900 prototype was the origin for Consolidated's Model 25. The two-seater fighter featured many novelties, such as a retractable undercarriage and enclosed cockpit. Although two prototypes crashed in the course of one week, 58 aircraft were ordered under designation Model 30, as P-30 high-altitude two-seater fighters (54 built) and A-11 attack planes (4 built).

A US Navy order led to the development of Model 28, an aerodynamically clean all-metal monoplane flying boat. It featured a parasol-mounted wing with retractable stabilising floats and leading edge-mounted radial engines. It made its maiden flight on 21 March 1935.

Centre left: PBY-5 variant, with its recognisable observation blisters.

Centre: PBY-1 of VP-6

Before war broke out, 209 examples of various subvariants were introduced into US Navy service as PBY. A further refined design, designated PBY-5, incorporated a new tail design, more powerful engines and larger fuel capacity. Distinctive observation blisters were introduced on the rear fuselage. It was built in large numbers, including the amphibious PBY-5A variant. Following an order from the British Royal Air Force in January 1940, the type was given the name Catalina. Further examples were ordered (or were received under lend-lease terms) by the US Navy, Canada, the Dutch East Indies, Australia, Brazil, the Soviet Union, and New Zealand. A final variant, PBY-6A, featured a further enlarged tail plane and was equipped with radar. The Catalina was the most built type of its kind, served in all war theatres and continued to fly in many post-war air forces.

Model 29 was developed as a four-engine patrol bomber and first flew on 17 December 1937 as XPB2Y-1. Development of the type progressed slowly, and after trials the tail surfaces were modified to improve stability. Production aircraft featured a redesigned nose, flight deck and fuselage spine. This, plus added wartime requirements, caused a weight increase resulting in the type being underpowered and unsatisfactory for its intended role as a patrol bomber. The aircraft, named Coronado following a British order, were converted into transport aircraft and served until the Japanese surrender. A commercial variant never progressed beyond the drawing board.

PB2Y-3, featured self-sealing fuel tanks and additional armour, but this made the type overweight. The type served as transport in the Pacific theatre.

PBY-6A amphibian, with AN/APS-3 radar dome above the cockpit.

Consolidated engineers were working on several large flying boat projects featuring efficient high-lift wings, but none were actually built. The Consolidator commercial Trans-Oceanic flying boat project was a family of concepts, dating from roughly August 1937 to July 1938. While the design progressed, it featured several wing variants. Another project was Model 30, a proposed successor for the Coronado featuring an improved tapered wing design. Development was progressing slowly due to commitment to other projects. Finally, a single example was ordered in April 1942 as XPB3Y-1. The project was cancelled before construction even began.

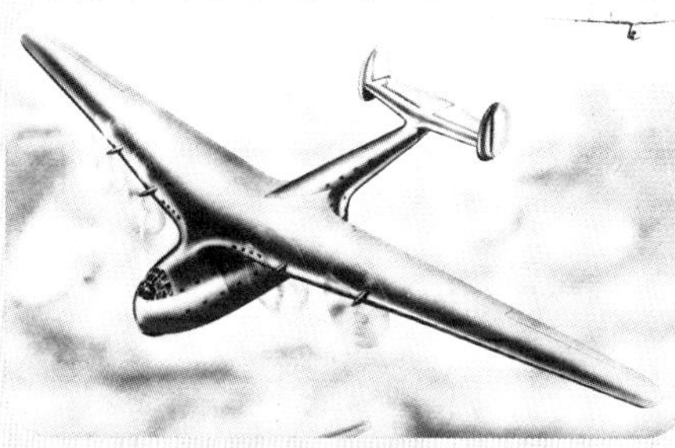

Consolidated 100-passenger flying boat project

VERY LONG-RANGE BOMBER

While the BLR types did not result in a suitable combat-ready aircraft, interest in the concept remained. The threat of war urged the USAAC to proceed with more vigour. A specification was drafted, in which high speed and long range were preferred over a heavy bomb load. The aircraft required pressurised crew compartments and needed to carry 2,000 pounds of bombs at a range of at least 5,333 miles at a speed of 400 mph. Experiences gained during the early months of the war in the air led to a revision of the specifications. More armour, defensive armament and self-sealing tanks were added to demands. The USAAC sent out Request for Data R-40B to a number of aircraft companies on 29 January 1940. Preliminary design studies were submitted on 8 April: Boeing model 345, Lockheed L-249, Douglas model 332, and Consolidated model 33. Martin followed shortly afterwards with Model 190. The designs were evaluated and designated by the USAAC as (in order of preference) XB-29 to XB-33. Negotiations for the delivery of two prototype aircraft and an option to purchase 200 production machines started. All companies were awarded contracts to produce wooden scale models for wind tunnel testing on 14 June. Out of the proposals, the Boeing, Consolidated and Martin designs were approved, and the companies received orders for the production of two prototypes. The Martin project, however, never progressed beyond the drawing board.

The Boeing XB-29 Superfortress prototype. (Collection: Mark Nankivil, via Nico Braas)

UTILISING THE DAVIS WING

During the summer of 1937, inventor David Davis was introduced to Reuben Fleet. Davis, who had no formal aeronautical education, claimed to have developed a highly efficient short- cord wing profile, which he had patented as "Fluid Foil" on 9 January 1934. Davis pitched his profile, hoping to sell a licence to Consolidated. Consolidated was working on a successor to the PBY and PB2Y types and the company's engineers were concentrated on efficient thin-wing designs. Despite initial scepticism, it was agreed to test Davis's wing profile in the wind tunnel of the Guggenheim Aeronautical Lab at Caltech in Pasadena. After a series of three tests, Davis design was found to be superior to Consolidated's design. The Davis wing (as it is now commonly known) showed a decrease in drag of 20%, which allowed for a substantial increase in speed, range and lift. Davis signed a contract with Consolidated on 10 February 1938, making him a consultant to the company.

The promising Model 31 remained a one-off.

The Davis wing design was implemented in Model 31, a twin-engine flying boat design, which was powered by two Wright R-3350 twin-row radial engines. The type first flew on 5 May 1939 and showed excellent performance. Testing showed that Model 31 was superior to both Consolidated PBY (Catalina) and Martin PBM (Mariner) designs. After the attack on Pearl Harbor the prototype was purchased by the US Navy which designated it XP4Y-1. An order for 200 was placed in October 1942, but this was later cancelled due to engine shortages.

The XB-24 showing off its clean lines.

The B-24H Liberator featured power turrets in the nose, dorsal and tail positions.

XB-24N during take-off.

The now-proven wing concept was used in Model 32, a new bomber project intended as successor to the B-17 Flying Fortress. A boxy fuselage was designed around two roomy bomb bays which could carry up to 9000 pounds. The Model 31's twin-tail arrangement was also incorporated into the new bomber. The four-engined bomber featured a tricycle landing gear and Fowler flaps in the wings, offering more lift during landing. The prototype was built in just over six months and took off for the first time on 29 December 1939, designated as XB-24. Even before the prototype had flown, two series totalling 45 machines were ordered. A total of 175 aircraft were ordered by France. The majority of these ended up in British service. Production was ramped up after the Japanese attack on Pearl Harbor. The B-24D was the first mass-produced variant with 2718 built, followed by 801 of the E variant. A power turret was introduced in the nose of the G series, which reduced the B-24's vulnerability to head on attacks. The B-24H, J, L and M variants were externally similar but differed by utilising various nose and tail turrets. With 14038 built, they were the most highly produced variants. To improve the stability of the B-24 a B-24D was modified with a single fin, which was designated XB-24K. Tests revealed that stability and control had indeed improved. In April 1944, the decision was made to continue production of the B-24 with a single tail assembly. In November 1944, a new Liberator variant emerged, with a lengthened nose, fitted with an Emerson 128 ball turret which improved speed and eliminated draft and buffeting caused by the earlier types of nose turrets. A new A-6D turret with an improved field of fire was placed in the tail. A contract for 5176 B-24Ns was signed but following the XB-24N, only seven YB-24Ns were built, before production of the Liberator ended abruptly on 31 May 1945.

The US Navy contacted Consolidated to develop an optimised version of the PB4Y-1 (essentially similar to B-24D) Liberator on 3 May 1943. The characteristic twin tails were exchanged for a large single-tail rudder and the fuselage in front of the wing was stretched by seven feet. The modified Liberator was known within the company as Model 40 and US Navy designation PB4Y-2 Privateer. The first of these made its maiden flight on 20 September 1943. A total of 1370 Privateers were ordered, but only 740 were completed when the war ended. Some 60 examples of a transport variant were built, designated by the US Navy as RY-3 and C-87C by the USAAF.

The PB4Y-2 featured a heavy defensive armament, allowing it to operate alone. It carried two dorsal turrets and a pair of fuselage blisters containing two machine guns, plus turrets in the nose and tail but no belly turret.

The Model 39, or R2Y Liberator Liner was an airliner derived from the B-24. It featured the bomber's wings and the single-tail configuration from the Privateer. The fuselage was a completely new circular cross-section design. Only one was built.

The first, completely unarmed, XB-32 prototype rumbling towards the runway.

DESIGN, EARLY DEVELOPMENT AND FIRST FLIGHT TESTING

Before receiving Request for Data R-40B, Consolidated had already worked on design study LB-25 for a land-based bomber powered by four Wright R-3350 radial engines. It featured a length of 77.7 feet and span of 135 feet. This study was first unveiled in March 1940. The XB-32 design evolved parallel to LB-25.

For the casual onlooker, the XB-32 may have looked like a B-24 on steroids, but it was certainly not. The XB-32 featured a far more refined, 83-foot-long cylindrical fuselage, housing two pressurised crew compartments. Eight crew members were seated in the front compartment. The bombardier sat in the nose of the aircraft, while the pilot, co-pilot, flight engineer and commanding officer were located in a roomy greenhouse cockpit. Directly behind them were workstations for the flight engineer, navigator and radio operator. In the aft of the front compartment, placed in a central position in the fuselage, two fire control officers operated two periscopic gun sights, connected to a Sperry P-4 fire control system. This system featured an analogue computer, which coordinated an impressive remotely controlled defensive armament. When the design entered into production, it was to be armed with remotely controlled dorsal and ventral turrets, fitted with four .50 calibre machine guns. When not in use, the turrets could be retracted into the fuselage.

The armament barbettes in the rear of the outer engine nacelles.

The retractable fuselage turrets caused heavy delays during production of the second prototype.

The rear of the outboard engine nacelles would house remotely controlled barbettes containing two .50 machine guns and a 20 mm cannon. A movable .50 machine gun would be fitted in the outer wing panels. These could be traversed by the fire control officers. The rear pressurised compartment housed two crew members, one operating a third periscopic sight. An extensively glazed tail dome allowed for a clear rearward view. From here, the two barbettes could be operated. The two pressurized cabins were split by a pair of bomb bays, but a pressurised tunnel ran over the top of the bomb bays. Similar to the B-24, the XB-32 prototypes featured two vertical tailplanes, but the horizontal stabiliser was constructed with a 7.5-degree dihedral. A distinctive feature of the XB-32 was the 135-foot cantilevered Davis wing. The centre section of the wing was built as a single piece with the fuselage centre section and housed six self-sealing fuel cells. The outboard wing panels housed the engine nacelles and a further three fuel cells per wing. The XB-32 was powered by four 2,200 hp Wright Cyclone R-3350 radial engines. The XB-32 featured a tricycle landing gear configuration. All dual-wheel landing gear struts were hydraulically retracted.

The 7.5-degree dihedral is clearly distinguishable in this photo.

production of two aircraft on 6 September 1940. This was later changed to three aircraft. The first prototype was to be delivered within 18 months after signing the contract, the second within 90 days of the first, and the third prototype within 90 days of the second machine. Wind tunnel tests were conducted on a 1/35th scale model by the USAAC Material Division at Wright Field, which revealed an unsatisfactory directional stability. A full-scale mock-up was reviewed and approved, after some modifications on 6 January 1941. Powerplant mock-ups were similarly reviewed and approved on 17 April. An additional contract for 13 YB-32 preproduction aircraft was signed in June.

41-141 during its test programme. Seen here taxiing with its bomb bay doors opened at Lindbergh Field. *(Photo via Nick Veronico)*

The first prototype carrying serial number 41-141 was ready on 1 September 1942, almost six months behind schedule. It was completed without the prospective defensive armament. At this stage, the type had received the name "Terminator". Production delays were caused by problems with the pressurised crew compartments. Availability of the engines caused further strains. After completing a programme of taxi tests, 41-141 made its first flight on 7 September 1942 from Lindberg Field in San Diego, California. The flight ended prematurely when severe tail buffeting was encountered, caused by a trim tab malfunction. There were more problems during the test programme. The process of examining and modifying the prototype was time-consuming, and new problems kept popping up. The engines and its turbo superchargers were giving a lot of trouble. A near-catastrophic in-flight fire caused by the extremely hot exhaust gases, was prevented just in time by making an emergency landing at Lindbergh Field. Although most of the problems were relatively minor, they caused a lot of delays, leading to the USAAC reconsidering, and finally cancelling the order for 13 YB-32 pre-production aircraft in February 1943.

The loss of Boeing's XB-29 bomber due to an engine fire on 18 February 1943 revealed that engine cooling on the B-32 needed several drastic modifications to prevent engine overheating. Consolidated hit a similar blow when, on 10 May 1943, 41-141 crashed just after take-off, injuring six crew members and killing the project test pilot Richard McMakin. The crash was a major setback for the B-32 programme as a lot of valuable instrumentation and test records were lost resumably caused by unspecified flap problems).
Investigation results on the XB-29's crash would benefit the XB-32 further test programme – the Consolidated programme did not have to endure further delays after implementing the necessary modifications.

The second prototype made its first flight on 2 July 1943. It still carries the roundel type without white bars. It ended its life during a fire drill.
(Collection: Pieto van Buysen)

41-142 with lowered Fowler flaps, which significantly increased the wing area during landing. *(USAAF photo)*

With the second XB-32 prototype number 41-142 production five weeks behind schedule, the future of the B-32 seemed very uncertain.

Externally, the second XB-32 prototype was more or less identical to the first prototype with pressurised cabin sections and the same double vertical tails, but with modified rudder tabs. Internally, the second prototype had been modified to improve the overall design. The defensive armament was introduced on the second prototype, which was reviewed with a fine-tooth comb. Despite the modern features, USAAC officials found the defensive firepower inadequate. The fuselage and engine nacelle mounted positions offered insufficient firepower, and all available gun positions left unacceptable blind spots – especially in the event of a head-on attack. The report suggested replacing the remote-controlled turrets with manned turrets and an increase of ammunition supply. The second prototype made its first flight on 2 July 1943 with test pilots Russel Rogers and Beryl Erickson at the controls.
Test flying revealed good flying characteristics, but the tail configuration kept causing stability problems.

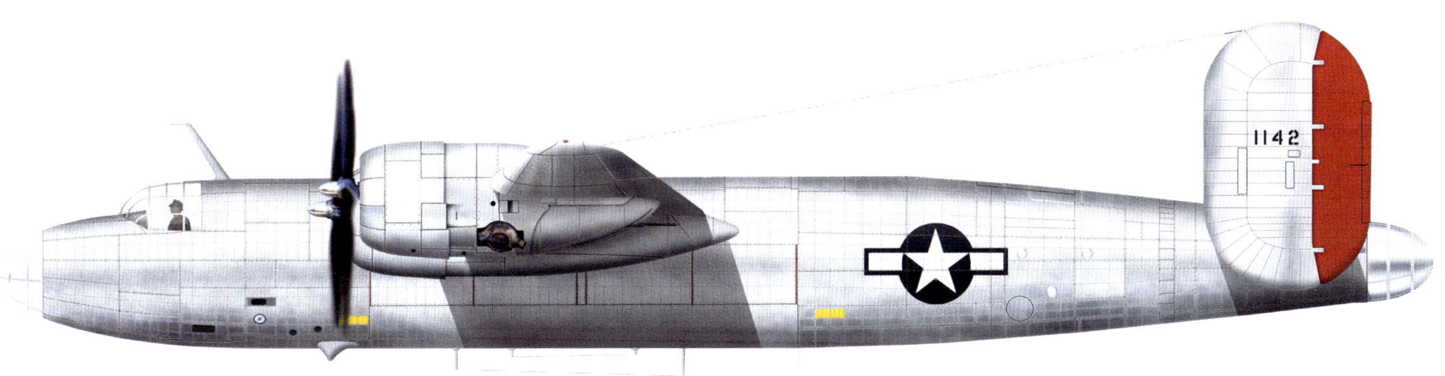

This side view reveals the tapering fuselage, giving the B-32 very aerodynamically clean lines.
(Colour profile by Srecko Bradic)

The XB-32 programme was inspected by USAAC engineers. Due to its delays, the XB-32 was no longer deemed fit for its original task. Furthermore, the XB-29 programme had progressed significantly. The USAAC inspection resulted in a lengthy report, concluding that the XB-32 was already obsolete to the 1943 requirements. Based on their report, the third prototype should be extensively modified, giving the XB-32 a tactical role, similar to the B-24. The pressurisation had to be deleted and defensive armament changed to manned turrets, including the installation of an Emmerson model 128 nose turret assembly. The nose of the aircraft should be extensively redesigned to allow better forward and side vision for the bombardier. The engine nacelles were to be modified, and the propellers exchanged for four-bladed Curtiss Electric units. Fuel, oil and bomb-release systems should be improved, and an automatic flight control system had to be installed. The design should be further optimised for easy maintenance and interchangeability of spare parts.

Cockpit overview of the first XB-32 prototype.

***Bottom:** Third prototype 41-18336, fitted with the single 16.5-foot-high tailplane.*
(Wright State University)

(Colour profile by Srecko Bradic)

The third prototype, in its definitive form, fitted with a dummy dorsal turret.

The third prototype received tail number 41-18336 and initially flew with the twin-fin tail. Stability problems were persistent and after 25 test flights, it was grounded in mid-1944. Many of the earlier suggested modifications were implemented during an extensive machine rebuild. The vertical tail planes were replaced by a single 16.5-foot-high B-29-styled tail. The third prototype made its first flight with this tail on 13 September. This solution was not satisfactory, and the tail was ultimately replaced by a single vertical fin similar to that of the Consolidated PB4Y-2 Privateer naval patrol bomber. The nose of the aircraft, up to the greenhouse-type cockpit canopy, was completely redesigned after the pressurised nose compartment was removed. The armament barbettes in the outboard engine nacelles were deleted, as were the projected wing guns. After the rebuild, active armament was tested on 41-18336. The third prototype was used to test several different manned turret variants.

Right: Several armament arrangements were studied in the Langley Memorial Aeronautical Laboratory. *(From: Naca MR No. L4L30a, www.ntrs.nasa.gov)*

PRODUCTION.

After the earlier cancellation of the contract for 13 YB-32 bombers, Consolidated was finally awarded a production order in March 1943. This was altered, after the modified third prototype had flown. Consolidated received an order for 300 machines, including a number of TB-32 aircraft intended for transitional training. The series production aircraft were designated Consolidated Model 34 and built at the company's Fort Worth, Texas site. The first production machine, 42-108471, was initially delivered with the 16.5-foot B-29 style tail, and later received the 19.5-foot unit. It was retained at the plant for use as a test aircraft and was continuously updated during its career.

MR No. L4L30a

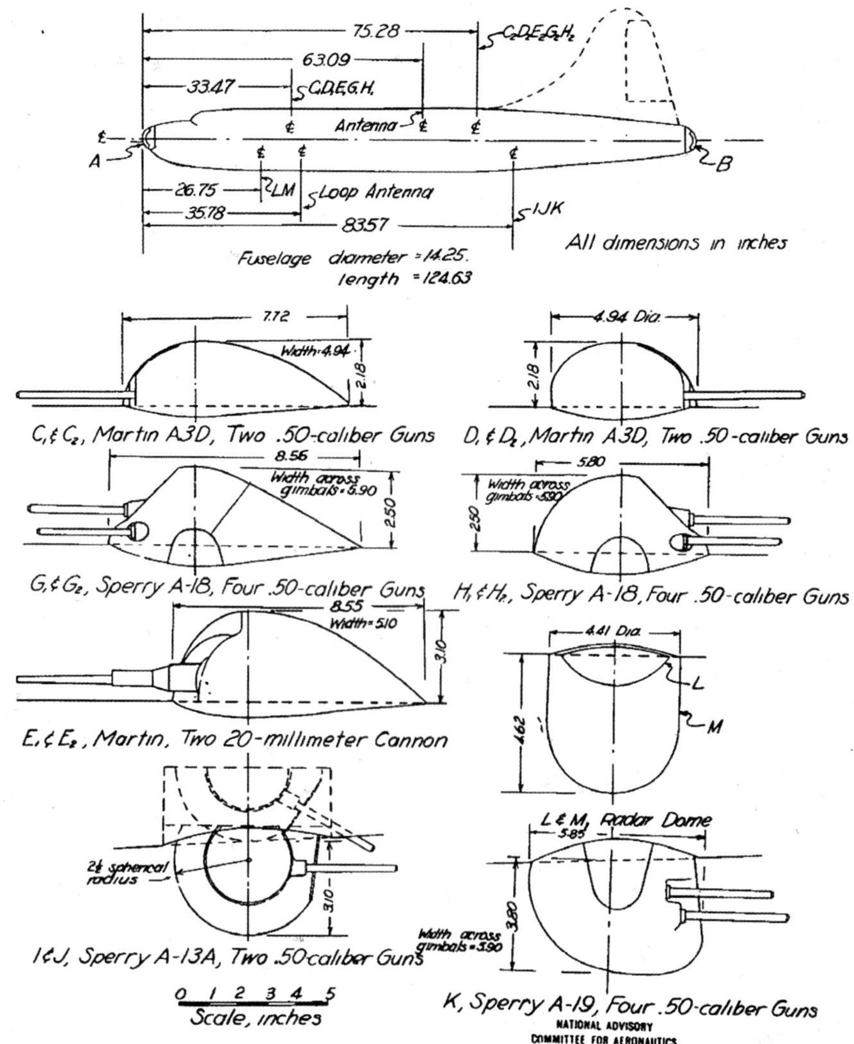

Figure 3 :- Location and general shape of turrets on a 1/8-scale model of the B-32 airplane.

Consolidated Vultee B-32 Dominator no. 42-108471 with early type B-29 vertical tail. *(Collection: Mark Nankivil, via Nico Braas)*

Front view of no. 42-108471. (USAAF photo)

42-108471, fitted with the definitive, Privateer-styled tail.

This view clearly shows the completely redesigned engine nacelles. A fixed AN/APQ-13 dome is fitted between the nose gear and bomb bay. (Collection: Thijs Postma)

Artist' impression of 42-108471. (Colour profile by Srecko Bradic)

The next machine, 108472, was the first to go airborne, on 5 August, but was lost in an accident during landing on 27 August. All following aircraft were delivered with the latter tail type, and several internal and minor modifications were implemented on the production line. The third production machine first flew in November. On the 21st of that month, an in-flight engine fire broke out, which initially went unnoticed by the crew. The flight was aborted, and an emergency landing was successfully made. Delivery of the next aircraft suffered delays due to a lack of 19.5-foot tailplanes.

42-108472 was the first production machine to fly but crashed shortly afterwards. (Collection: Phil Flaschberger, via Nico Braas)

By the end of December, only five aircraft had been delivered to test centres. To make matters worse, a lot of complaints concern-

The TB-32 variant was unarmed. The nose, tail and fuselage turrets were faired over, and all bombing and radar equipment was removed. The bomb bays were retained, however, and 750 pounds of ballast was added to compensate.

ing mechanical failures and poor workmanship were received. Despite the very poor outlook, further orders were received, totalling over 1700 aircraft.

Production aircraft were fitted with four-bladed Curtiss Electric constant speed propellers, which had a diameter of 16 feet 8 inches - the largest up to that moment. The two inboard propellers could be reversed to a maximum of -15.8 degrees, which gave the B-32 the novelty of thrust reversal. This shortened the landing roll and allowed the crew to reverse the aircraft during ground manoeuvring. The B-32s were armed with ten Browning .50 machine guns. A Sperry A-17 turret was fitted in the nose and tail of the aircraft. Two Martin A-3F-A dorsal turrets were located on the top of the fuselage. They featured a Plexiglas fairing behind the turret, to maximise streamlining. A retractable Sperry A-13 was placed behind the two bomb bays. All turrets were equipped with two machine guns. The B-32 could carry a maximum bomb load of 20,000 pounds. The B-32 could carry various bomb loads, including light case 4000-pound bombs.

Similar to the B-24, the bomb bay doors were of the roll-up type.

Above: *Series production of fuselage fronts.* *(The Portal to Texas History, Lockheed Martin Aeronautics Company, Fort Worth)*

Below: *A completed fuselage, with wing centre section is ready to be lowered onto an assembly scaffolding.* *(The Portal to Texas History, Lockheed Martin Aeronautics Company, Fort Worth)*

PRODUCTION OVERVIEW

B-32-1-CF – 42-108471 to 480
Flight testing aircraft. Wright R-3350-23 engines. Radar bombing equipment (AN/APQ-5B developed for low altitude bombing, and AN/APQ-13, an improved version of the H2X ground scanning radar) and long-range navigation equipment were installed. 10 built.

B-32-5-CF – 42-108481 to 484
Flight testing aircraft. Twin rudder tabs were made standard. 4 built.

TB-32-5-CF - 42-108485 to 495
Trainer aircraft, 8000 pounds lighter compared to their combat-equipped counterpart. 11 built.

Female employees working on a Wright R-3350 engine. Early development of the engine was troubled, but after the war it powered many types, such as the Lockheed Constellation and Fairchild C-119.
(The Portal to Texas History, Lockheed Martin Aeronautics Company, Fort Worth)

The engine power-egg was completely interchangeable.
(The Portal to Texas History, Lockheed Martin Aeronautics Company, Fort Worth)

Right: *Placing the central segment of the tail on the 48th production machine, a TB-32-10-CF version.*
(The Portal to Texas History, Lockheed Martin Aeronautics Company, Fort Worth)

Although the trainer aircraft were unarmed, the fuselage featured the circular openings for dorsal turrets.
(The Portal to Texas History, Lockheed Martin Aeronautics Company, Fort Worth)

Artist's impression of TB-32 42-108495. (Colour profile by Srecko Bradic)

TB-32-10-CF - 42-108496 to 520
Redesigned bombardier's entrance door, replacement of SCR-269-G radio compass with AN/ARN-7 set, and installation of engine fire extinguishers. 25 built.

TB-32-15-CF – 42-108521 to 524
Introduction of de-icer boots. 4 built.

B-32-20-CF – 42-108525, 42-108526, 42-108528 to 545
Combat-equipped aircraft. Scanning blister installed in rear fuselage for observing engine nacelles and landing gear. Bomb bay modifications, including the option to carry cargo. Life raft stowage compartments on top of the fuselage. 20 built.

B-32-21-CF – 42-108527
Converted to paratrooper transport. All bombing equipment was removed, and benches were installed in the rear bomb bay and rear fuselage. Remained at the factory. 1 built.

B-32-25-CF - 42-108546 to 570
Modified fuel system to allow auxiliary tanks in the bomb bay. AN/APN-9 LORAN. 25 built.

TB-32s at the final stage of the assembly line. (Loc.gov)

The completed bombers were placed on transport cradles to move the large aircraft diagonally through the assembly hall doors.
(The Portal to Texas History, Lockheed Martin Aeronautics Company, Fort Worth)

TB-32s above a Texan landscape. The front plane, 42-108524, is one of the final batch of four TB-32 aircraft, which were fitted with de-icer boots.

Only 45 fully combat-equipped aircraft were delivered before production was terminated. Here, 547 is fitted with a retractable AN/APQ-13 dome.

B-32-30-CF - 42-108571 to 577
Introduction of stabilised Sperry A-17A nose turret, installation of countermeasure equipment (AN/APQ-2, AN/APT-1 and AN/APT-2) and improved APQ-13A bombing radar equipment. Strengthened wing trailing edge. 7 built, and the last three aircraft were flown directly to storage and scrapped.

B-32-35-CF - 42-108578 to 584
Increased ammunition; flown directly to storage and scrapped. 7 built.

B-32-20-CO 44-904486 to 44-904488
Same as B-32-20CF but assembled by the Consolidated San Diego plant. One aircraft was accepted with the remaining two units flown directly to storage and scrapped. 3 built.

Based on the above, 117 aircraft had been built when production was terminated. Three prototypes had been built before series production commenced. One aircraft was lost in an accident before it could be delivered, and ten were immediately put in storage and scrapped. 106 aircraft were delivered: 66 combat-equipped aircraft and 40 trainer versions. An unknown number of aircraft were still under construction. These were stripped of all equipment and engines and were scrapped.

TESTING AND TRANSITIONAL TRAINING

When the B-29 finally saw action on 5 June 1944, the B-32 programme was severely lagging behind. It was not until 15 August 1944 when Material Command at Wright Field issued a directive for testing of the B-32 during a 200-hour programme. Investments in the B-32 had been too large to simply terminate the programme, so testing work continued. During this period the official name for the B-32 was changed from "Terminator" to "Dominator".

The 100th production machine being moved across the production facility. It is fitted with a retractable AN/APQ 13 dome. *(The Portal to Texas History, Lockheed Martin Aeronautics Company, Fort Worth)*

FW 101, 42-108571, being weighed prior to exiting the production facility.
(The Portal to Texas History, Lockheed Martin Aeronautics Company, Fort Worth)

A Stinson L-5 Sentinel liaison aircraft is dwarfed by the TB-32. Note the opened bombardier's entrance door in the nose of the Dominator.

The planned test programme was postponed due to production delays. The first ten aircraft were assigned for test duties, but by May 1945 only six aircraft had become available. ATSC would receive the following aircraft: 42-108477, 484, 478, 481, 482, 525, 526, 533, 535, 540, 541, 542, 546, 547, 571, 572, 573, 574, 575, 576. AACTC received 42-108480, 534, 535, 536, 537 and 538.

Release of triple-suspended AN-M30 100 lb GP bombs by 42-108547, which received Eglin Field number 591 when it served with Squadron "E", 611th AAFBU, Eglin Field, Florida.
The Field numbers did not correspond to the tail numbers:
580 = 42-108484,
581 = 42-108477,
589 = 42-108535,
591 = 42-108547,
596 = 42-108574
(Photo via Nick Veronico)

The test programme revealed several major problems, which included a high risk of engine fires and extremely high flight deck noise levels. Other issues that needed to be modified were instrument and control layout and sub-systems inadequacies. On several occasions, poor craftsmanship and faulty construction methods came to light. All issues required redevelopment and modification, which kept test aircraft grounded for unusually long periods. All problems were rooted out of the subsequent build tranches. Overall build quality improved after quality control was improved. On the plus side, the B-32 was judged positively. It had excellent low-speed control, good take-off and landing qualities and quick control response. The B-32 was a stable bombing platform, and its defensive armament provided good all-round protection. Ground manoeuvrability was unmatched. Testing continued until October 1945, the moment the B-32 was cancelled.

TB-32 Dominator numbers 42-108490 and 42-108495 photographed at Fort Worth, February 3, 1945. The aircraft carry Station Identification Code OM on the nose, followed by a number. (Photo via Nick Veronico)

The sixth production B-32, which participated in the B-32 service tests. (Collection: Thijs Postma)

TB-32 Dominator no. 42-108517 with Field number OM31.

Consolidated TB-32 Dominator 42-108495, with FW No 25 stencilled on the nose.
(Collection: Thijs Postma)

Training of new crews for the B-32 was seriously restricted by the number of available aircraft. The 2519th Army Air Force Base Unit at Fort Worth was responsible for testing and training B-32 pilots and complete crews. The 2519th offered the first transitional training of crew and the final phase of training. Most of these men were experienced B-24 crews. 57 aircraft, both B-32 and TB-32 variants, were assigned to the unit and delivered between January and July 1945. Several incidents with main gear collapses hindered the training schedule. All aircraft were grounded in May 1945 after a number of such accidents. Flying continued in July after a modification programme.

The progress of the training programme was often frustrated by a lack of spare parts. Gear collapses and engine fires put aircraft out of commission or would lead to write-offs. On numerous occasions, spare parts had to be made by hand because regular production had already been stopped. Ground crews were trained at technical schools located at Keesler Field (Mississippi), Lindbergh Field (near San Diego) and Chanute Field (Illinois). The technical schools received spare parts, including complete landing gear units, engines and one complete airframe (42-108489).

Losses during testing and training		
Aircraft	**Date**	**-**
42-108472	27 August 1945	Written off after landing gear door issues
42-108473	05 March 1945	Written off after accident – a Lockheed C-60 crashed into the hangar housing the B-32
42-108495	08 March 1945	Crashed after engine fire
?	10 March 1945	Crashed after engine fire
42-108475	23 March 1945	Crew bailed out after mid-air fire

Consolidated TB-32 42-108489 while used by the Technical School at Keesler Field. *(Doug Olsen collection, via Nick Veronico)*

OPERATIONAL

The Air Corps had already prepared detailed war plans during August 1941, should the US get involved in the war. In these plans, the B-32 was destined to be part of a large, very heavy bomber force, required for a precision bombing campaign on Germany's industrial heart. Neither the B-29 nor B-32 would be available in time for the war in the European theatre. Both types suffered delays. The B-32 was originally planned to be introduced into service with several groups in the Mediterranean Theatre of Operations. It would replace the B-24 then operating with the 8th and 15th Air Forces, giving these groups a long-range striking capability.

Bottom: *The Lady is Fresh, still with an incomplete nose art, runs up its engines.* *(Loc.gov)*

By the spring of 1945, the B-32 was finally declared combat-ready, but service in Europe was no longer necessary when the war entered the final stages in April 1945. It was none other than Lt. Gen. George Kenney, commander of the Far East Air Forces (FEAF), who secured a combat role for the B-32. Kenney needed long-range bombers to bomb the oil refineries in the Dutch East Indies, which were vital to Japan's war effort. After a lengthy campaign to secure B-29s for this task, Kenney switched his attention to the B-32, and with success. After two weeks of debates and demonstrations, it was agreed to continue B-32 production, with the goal of replacing the B-24 in the Pacific. A number of B-32s were assigned to the FEAF for combat tests. A test detachment was formed, which would be commanded by Colonel Frank R. Cook of the ATSC (Air Technical Service Command). Cook had previously been part of the B-29 project.

The 33-man test detachment travelled to Consolidated' Fort Worth plant, where three B-32s had been prepared for deployment to the Philippines. The three machines were of the B-32-20-CF variant, the first batch of combat-equipped aircraft:

42-108529 The Lady is Fresh
42-108531 (unnamed)
42-108532 Hobo Queen II

The assigned crews made three flights to familiarise themselves with their new aircraft. On 3 May, 42-108531 suffered a landing gear issue and ground looped during landing. It was heavily damaged, and a new aircraft was assigned: 42-108528, which had previously served as an armament test aircraft. It had to be prepared for its new task and inspections and repair work delayed departure by a week. The group would first depart for Mather Field, California. On the 11th, the first leg of the transfer flight was made by 532, followed by 529 the following day. 528 left on 13 May and had a troublesome flight. It had various malfunctions, and the landing gear had to be cranked down manually. To make matters worse, all spare parts, which were to be shipped by air, ended up in Biak, New Guinea, according to an earlier flight plan. The route had been changed several times, which may have caused the mistake. In order to repair the 528, spare parts were collected at the Consolidated plant and were to be transported in specially crafted cargo bins, which were hung in the bomb bays.

Hobo Queen II, with only a name written on the starboard side of the nose, during its transfer flight to the Philippines.

A Consolidated F-7A Liberator of the 20th Combat Mapping Squadron, photographed at Clark Field. The newly arrived B-32s are visible in the background. (Photo via Srecko Bradic)

529 and 532 departed on 15 May to Honolulu, Hawaii. Several issues had to be solved on 532, while 529 needed an engine change. 528 was finally flown to Hawaii on the 17th, but again technical issues occurred, this time on Loran, with compass and radar equipment. 528 was able to catch up with the other two Dominators on the 22nd, when it landed at Harmon Field, Guam. After repair and maintenance work, the three aircraft were ready for the final leg of the journey, to Clark Field in the Philippines. The first two machines departed on the 24th, while 528 remained an extra day for repairs. Once more, the crew of 528 had to deal with malfunctions, including engine trouble and a leaking fuel pump. After a total of 175 flying hours the Dominators reached the Philippines. Along the way, the crew had to overcome technical issues of all sorts, caused by mechanical and electrical failures, engine trouble, leakages, and construction faults.

The 386th (Light) Bomb Squadron operated from Floridablanca airfield, a former Japanese base. A pair of asphalt runways had been completed just before the arrival of the B-32s. The squadron had operated the Douglas A-20 Havoc light bomber but handed these over to 388th and 389th, sister squadrons of the 312th Bomb Group. Lt. Col. Selmon Wells was made commander of the combat test. The objective of the test was to determine if the B-32 was a suitable combat aircraft, which included aspects such as serviceability and effectiveness of all armaments and equipment.

The first B-32 mission was against Japanese positions in the town of Antatet, Northern Luzon. All three Dominators would participate, armed with nine 1,000-pound bombs. The 528 suffered engine trouble and had to abort take-off. The other two successfully bombed the target from 10,000 feet and took photographs of the results. The mission was a success, and the next day, 386th was redesignated "Very Heavy Bomb Squadron" and transferred to Floridablanca for training new crews. 528 remained grounded and required an engine change. During the next days, training flights were made, mostly with The Lady is Fresh and Hobo Queen II.

Hobo Queen II seen shortly after arrival at Clark Field. A bikini-clad girl is painted under the nose. (Loc.Gov)

Mission 1		
Aircraft	**Date**	**Objective**
528 529 532	29 May 1945	Japanese positions, Antatet, Luzon

Mission 2		
Aircraft	**Date**	**Objective**
529 532	2 June 1945	Basco Airfield, Luzon

The Lady is Fresh, parked on a separate space is inspected by 312th BG personnel. (Loc.Gov)

532, named Hobo Queen II. The photo is part of a series taken at Clark Field. *(loc.gov)*

Basco airfield, on Batan Island, was attacked during the morning of 12 June. Each bomber, armed with forty 500-pound HE bombs, individually made two bomb runs each from a height of 16,000 feet. Photo assessment showed that the airfield was rendered inoperable. During an inspection of 529 (The Lady is Fresh) shrapnel damage was found in a stabiliser, which must have been received during the previous weeks training flights.

Two Dominators returned to Formosa the next morning to bomb a sugar mill (alcohol refinery) at Taito (Taitung), located on the eastern coast of Formosa. The first bomber dropped its eight 2,000-pound bombs from 15,000 feet but missed by a wide margin. The second crew bombed more accurately, hitting warehouses and part of the refinery. Some inaccurate anti-aircraft fire was encountered when the second plane was over the target.

All three B-32s participated in a raid on the city of Taito and dropped a deadly load of one-hundred-and-twenty 500-pound incendiary cluster bombs from 19,000 feet. The bomb bay doors on one of the machines jammed and had to be opened using an emergency release. During the attack the aircraft dropped aluminium strips called Rope, to disrupt the Japanese radar-guided anti-aircraft guns. Some heavy anti-aircraft fire was encountered. The attack was a success, Taito was left behind burning heavily.

Mission 3		
Aircraft	**Date**	**Objective**
528 532	13 June 1945	Koshun airfield, Formosa

Two B-32s took off to bomb Koshun airfield, located near the south-eastern coast of Formosa (Taiwan). It was the first time the B-32s crossed the Philippine Sea, a 960-mile return trip. Each bomber was loaded with twelve 1,000-pound bombs. The two planes made separate bomb runs from 12,000 feet. Three bombs hung up on 528, which were later jettisoned over sea. The attack was unopposed by fighters or anti-aircraft fire and after 6:25 hours both aircraft landed safely at Floridablanca.

Mission 4		
Aircraft	**Date**	**Objective**
529 532	15 June 1945	Sugar mill, Taito, Formosa

Mission 5		
Aircraft	**Date**	**Objective**
528 529 532	16 June 1945	Taito town, Formosa

The three Dominators of the test department at Clark Field. (*Norbert Crane collection, via Nico Braas*)

532 took off at 19:00 hours for a night reconnaissance mission, searching for Japanese shipping. It was loaded with nine 500-pound bombs and an extra fuel tank in one of the bomb bays. After a four-hour flight, it arrived over the designated patrol area, but found no ships. At 02:38 the secondary target, Haikow town, on the northern coast of Hainan Island, was bombed. During the mission, the crew sent hourly weather reports to base in order to test the communication equipment.

Mission 6		
Aircraft	**Date**	**Objective**
532	17/18 June 1945	Anti-shipping sweep Tonkin gulf

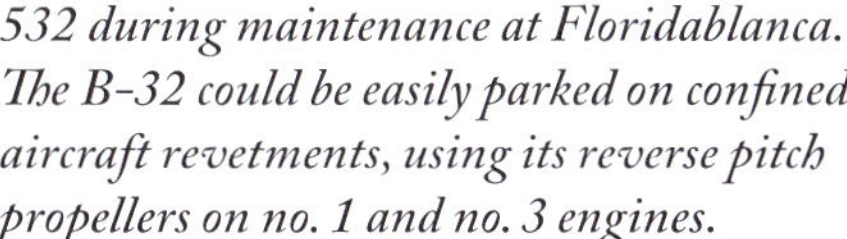

532 during maintenance at Floridablanca. The B-32 could be easily parked on confined aircraft revetments, using its reverse pitch propellers on no. 1 and no. 3 engines.

The objective for the seventh mission was three railroad bridges along the east coast of Formosa. All three aircraft were prepared. 528 and 532 (Hobo Queen II) carried twelve 1,000-pound bombs, while 529 carried nine. The bombers made individual bomb runs, dropping three bombs per run. The bridges were straddled with near hits, but none were confirmed as destroyed. 528 had to feather a propeller during the return trip to Floridablanca.

Mission 7		
Aircraft	**Date**	**Objective**
528 529 532	19 June 1945	Railroad bridges, east coast of Formosa

Two bombers were loaded with four 2,000 bombs each for an attack on the railroad yards near Suo, on the north-eastern coast of Formosa. While approaching the target, the crew had to divert to the secondary target due to poor visibility. Only the first crew were able to drop their payload in their given target, Paiyapai Bridge north of Taito. The second crew headed for a reserve target: warehouses in the western part of Tamari. These were demolished with direct hits.

Mission 8		
Aircraft	**Date**	**Objective**
528 529	20 June 1945	Railroad yards, Suo, Formosa

Mission 9		
Aircraft	**Date**	**Objective**
528 529	22 June 1945	Butanol refinery, Heito, Formosa

The ninth mission entailed the destruction of a sugar refinery (producing butanol for the aviation fuel industry) and nearby heavy anti-aircraft artillery positions. For the first target, 528 was loaded with forty 500-pound bombs and 529 with seventy-eight 260-pound fragmentation bombs. 529 was the first to attack. Its fragmentation bombs missed the artillery positions but did hit barracks and other military buildings. 528 followed and bombed its target from an altitude of 15,000 feet. Thirty bombs detonated in the target area, but 10 were hang-ups and were jettisoned as quickly as possible. Both aircraft received heavy anti-aircraft artillery fire and 528 was damaged in an engine nacelle and trailing edge of a wing.

Mission 10		
Aircraft	**Date**	**Objective**
532	23/24 June 1945	Enemy shipping, Canton River mouth, China

Mission 11		
Aircraft	**Date**	**Objective**
529 532	25 June 1945	Railroad bridges, Kiirun, Formosa

A single aircraft was sent out on a night shipping search in the Canton River mouth (also known as Zhujiang River or Pearl River). No sightings were made inside the given target area, so the secondary target, Sanchau Airfield, was bombed from 10,000 feet with nine 5,000- pound bombs using H2X radar. No results were observed due to cloud cover. Three bombs failed to release and were jettisoned over sea. During inspection, one of the rear bomb bay doors was found to be damaged during the bomb drop.

Two Dominators were prepared for the final test mission, a heavily defended railroad bridge north of Kiirun. It was the longest flight during the combat test. 532 was able to drop its nine 1,000-pound bombs over the area from an altitude of 20,000 feet. All bombs fell wide. 529 had to divert to the secondary target, the town of Giren. All bombs struck within the target

Bottom: The B-32 was first displayed to the public on Air Force Day, 1 August 1945.

B-32 Dominator no. 42-108529 wearing the first variant of the 'The Lady Is Fresh' art. (Collection Phil Flaschberger, via Nico Braas)

area. No enemy anti-aircraft artillery was encountered, and no enemy aircraft were seen.

After the test missions, an assessment was written. Twenty-three sorties had been flown, of which one had to be aborted. 133,27 tons of bombs were dropped and 140:36 combat hours were made. In general, the written reports were favourable. There had been no critical malfunctions and most technical issues (such as landing gear retraction and oil leaks) were common problems during operational service.
The B-32 had proven itself as an aircraft without vices, with excellent landing and formation flying characteristics. Pilots praised its responsiveness to the controls. Crews could easily convert to the type and navigators positively remarked about their roomy accommodations. Bombardiers, however, felt that the nose area of the Dominator was cramped and visibility was poor. The B-32 was also significantly faster, could fly further and carry a larger bomb load compared to the B-24. There were also issues, concerning engine overheating, and warping of the airframe structure, which caused poorly closing inspection hatches and entry doors. On some occasions the bomb bay doors did not operate correctly, and ammunition feed for the nose and tail turrets malfunctioned. The report concluded that none of the issues were critical, or unsolvable, and as a whole, the B-32s performance and serviceability had improved during the test period.

The new artwork on The Lady is Fresh, which was painted during the second half of June 1945.

578 on the ramp at Yontan Airfield.

Hobo Queen II during maintenance on engine no. 1 (Loc.Gov.)

After the completion of the test missions, the three Dominators were used for training new crews. The 387th squadron received orders to turn in its A-20 bombers and prepare for B-32 training on 4 July.

Although the test phase had been completed, the B-32s of 312 Bombardment Group remained operationally active. On 6 July, all three bombers were involved in an attack on a sugar refinery in Tako, Formosa. 528 and 529 carried twelve 1,000-pound bombs and 532 carried nine 1,000-pound bombs. The results were poor only six out of these 33 bombs hit the target. On the 13th, a single B-32 took off for a night mission, but had to return due to poor weather conditions.

On 23 July, the 386th and 387th Bombardment Squadrons received orders to transfer to Yontan Airfield, Okinawa.

On the 30th, B-32 no 530 arrived at Florida US, and 14 more were expected to arrive in the near future. Meanwhile, the journey to Okinawa was prepared. The ground crews and equipment travelled ahead by ship, leaving on 6 August, the same day as the bombing of Hiroshima by B-29 Enola Gay. 528 and 532 arrived at Yontan on the 11th; 530 and the newly assigned 543, 539 and 544 arrived the next day. 578 landed on the 13th and 529 could be flown over after an engine change on the 21st. The 531 was the final B-32 to arrive on 4 September. Of the newly assigned aircraft, one was adorned with nose art: 543 was named Harriet's Chariot.

The landing ships that transported the ground crew and equipment arrived on 12 August and set up camp at Yontan Airfield. During the night of 13 to 14 August, two missions were flown. 528 was sent out on a shipping search over the Chinese Sea. After completing the mission, an engine caught fire during landing, taking 528 out of commission. Later, two Dominators flew an armed reconnaissance. The first, loaded with nine 500-pound bombs, patrolled along the Korean coast but did not find any target. The second bomber attacked a 75-foot sloop, scoring two near misses. The ship was seen listing, with the crew abandoning ship. A second 150-foot ship was attacked with machine gun fire. On the next day two B-32s were sent out on a reconnaissance flight between Korea and Honshu, but the aircraft were radioed with an order to return to base: the Japanese had finally agreed to surrender.

Although the Japanese had now formally surrendered, 386th Bombardment Squadron continued operations. The Japanese had not responded to a request to send a report on the condition of airfields in the Tokyo area, so B-32s were sent to gather intel. It was now given a number of reconnaissance missions, to observe whether the Japanese were acting in accordance with the ceasefire. On 16 August, 532 and 543 made a surveillance flight to the airfields around Tokyo. Despite the ceasefire the aircraft was met with some inaccurate anti-aircraft artillery fire. Ten enemy fighters intercepted the B-32s, of which two were claimed shot down. 543 returned to Yontan with damage to an engine.

539, pushed aside after sustaining battle damage on 17 August.

On the next day 532, 539, 543 and 578 took off early in the morning for another photo reconnaissance mission. They had to make cartography photo runs over northeast Tokyo, flying two miles apart. During this initial phase, one of the B-32s was intercepted by a fighter identified as a Nakajima Ki-44 Shoki (Allied code name Tojo), which at first flew along the Dominator, but then attacked from two o'clock. The fighter was chased away by the gunners. Due to cloud cover, the four B-32s diverted to photographing their secondary targets, a number of airfields. 543 was fired at by heavy anti-aircraft artillery, which damaged an engine nacelle and port wing. While Harriet's Chariot continued its mission, it was intercepted by ten Tojo fighters, which made indi-

543 was in action for the first time on 16 August. (Loc.Gov.)

578 Taxiing towards the runway on Yontan Airstrip before take-off, 25 August 1945. (Loc.Gov.)

vidual attacks. One was damaged by the nose gunner, and a second was hit by the top gunner and disengaged, tracing black smoke. The remaining Japanese fighters continued to give chase but remained at a safe distance. 543 received hits in both wings until the Japanese finally broke off their pursuit. 539 was damaged in a similar fight. It had to feather one of the engines and was hit in an aileron and a wing flap. 578 was also attacked; the Japanese pilot made a head-on attack, passed below the B-32, and made a second stern attack. When it passed the bomber, the upper rear turret gunner placed hits in the engine and cockpit area of the enemy fighter, which escaped into cloud and was last seen diving away while on fire. It was claimed as shot down. After all Japanese fighters had disappeared, the B-32s finally headed home. Two aircraft were damaged and 539 had to be written off due to a lack of spare parts. It was cannibalised for parts and never flew again.

Only after the war, it was found that famous Japanese ace Saburo Sakai was among the intercepting pilots. It was his last action of the war.

Four B-32s were made ready for another photoreconnaissance of the Tokyo area on Saturday 18th August. A single Consolidated F-7B* flew ahead. 532 was joined by the 543, 544 and 578 and headed towards Japan. Both 543 and 544 developed engine trouble and had to return to base after five hours of flying. The remaining aircraft were almost immediately picked up by Japanese radar, and quickly received intensive anti-aircraft fire. After joining each other at an altitude of about 10,000 feet, the bombers were jumped on by a group of 14 Japanese fighters. Japanese ace Sadamu Komachi recalled after the war that the Japanese could not bear the sight of the American bombers flying unopposed over Japan. Three of them made individual attacks on the 532, which were fended off by the gunners. One fighter broke off, trailing smoke. Meanwhile, 578 was targeted by the other fighters. One of the engines was hit, which had to be feathered. During a frantic dogfight, the Japanese kept attacking individually. The nose turret gunner hit one of the fighters, as did the tail gunner. The latter scored a confirmed victory when the attacking fighter burst into flame and exploded. More attacks followed, one of which hit the B-32 from the side. Two crewmen in the rear compartment were wounded critically. 19- year-old Sgt. Anthony Marchione from Pottstown, Pennsylvania would succumb to his wounds. During the ensuing battle, one more fighter was claimed to have been destroyed by the rear upper turret gunner. His station was hit with cannon fire, however, disabling the guns and knocking the gunner unconscious. The pilot put his Dominator in a steep dive in order to hide in thick cloud some 1,500 feet below. Finally, the Japanese retreated.

* A modified Liberator fitted with five photo cameras

A graphic image of a B-32 with no. 3 engine emitting smoke.
(Collection: Phil Flaschberger via Nico Braas)

The three, four-engine bombers returned to base when the single F-7 was hit by anti-aircraft artillery. After an 11.5-hour flight, the bombers landed at Yontan. 578 landed on three engines, with one dead and two wounded crew. Two enemy aircraft were shot down with a third probable. They were the last enemy aircraft downed by a US warplane, and Marchione was the last AAF airman to be killed in action during World War Two.

On the 25th, a new mission to the Tokyo area was thwarted by malfunction and poor weather. Two out of four machines aborted the flight due to technical issues, and the remaining two Dominators turned around when approaching Japan, due to a tropical storm over the area. Two days later, another mission was cut short when both aircraft suffered landing gear retraction problems.

On 28 August, five B-32s would participate in what would be the last mission of the war, a communications flight over Atsugi and Tokyo. The final Dominator to take off was 544, but an engine lost power during take-off. An attempt was made to abort, but it was too late; the B-32 dropped back on the ground at the end of the runway and burnt out fiercely. All 13 men aboard were killed. 529 had to turn around with landing gear issues. The remaining three bombers continued their mission. 528 developed engine troubles in two engines, forcing the crew to bail out. All men bailed out successfully, but only 12 were rescued by US Navy vessels. Corporal Morris Morgan was never found. Sergeant George Murphy died shortly after his rescue.

529 had made 285 ¾ flight hours. It was damaged during a typhoon during the night of 9/10 October, and would be the last Dominator to leave Okinawa, on 16 October.

Bottom: *530 painted with the slogan "Direct from Tokyo", shortly before its trip to the US.*

530 was prepared for a flight to New York, carrying photos of the Japanese surrender ceremony. For this flight, it received a large 5th Air Force Patch below the cockpit with the slogan *Direct from Tokyo*. A 312th Group insignia was painted on the nose. The plane left on 31 August, following the route Kwajalein, Honolulu, Mather (California).

After the end of the war in the Pacific the B-32 was no longer necessary and further use was limited to decommissioning the type. The B-32 programme was formally terminated on 12 October 1945. All production orders had already been cancelled by June, and construction of aircraft which were still on the line was halted on 18 September.

On 1 September, there were seven airworthy B-32s left on Okinawa. Of these, 532 was damaged due to a nose gear malfunction on 9 September and was declared a total loss and 529 was damaged in a tropical storm. The remaining B-32s were flown to the United States, departing on 14 and 15 October. After arrival, they were flown to Kingman disposal in Arizona to be scrapped.

All flyable B-32s in the US were ordered to be flown to storage and disposal sites. Six B-32s were flown from the production line straight to Walnut Ridge disposal site. Some 50 B-32s still on the line at Fort Worth were chopped up and shipped to smelters. By 1947 only five B-32s remained in storage at Davis-Monthan, Tucson (Arizona). Amazingly, two airframes survived into the 1950s. 42-108489 survived the scrapping programme and remained at Keesler Field for training mechanics. It was eventually was used for firefighting instruction in 1952. 41-18336 – the third prototype – ended up at McLellan AFB and similarly was used for firefighting training until 1956.

The Hobo Queen II received a new, much larger pin-up painting, placed further back on the fuselage. The bikini was exchanged for a more daring outfit… it is seen here without armament, and with 10 mission markings on the nose. The nose area is rippled after a landing gear failure on 9 September. *(Norbert Crane collection, via Nico Braas)*

OM10 (42-108497), OM12 (42-108488) and OM20 (42-108500) waiting for their turn in the smelter….
(Davis-Monthan, 1947. Photo via Nick Veronico)

42-108480 at Kingman Disposal Field.

Several dozens of B-32s and TB-32s were gathered at Walnut Ridge to be scrapped.

42-108474, the fourth production aircraft, originally named Flaming Mamie, was initially reserved for display at the Air Force Museum but was eventually scrapped at Davis-Monthan in August 1949.

Cockpit overview of a TB-32. (Photo via Srecko Bradic)

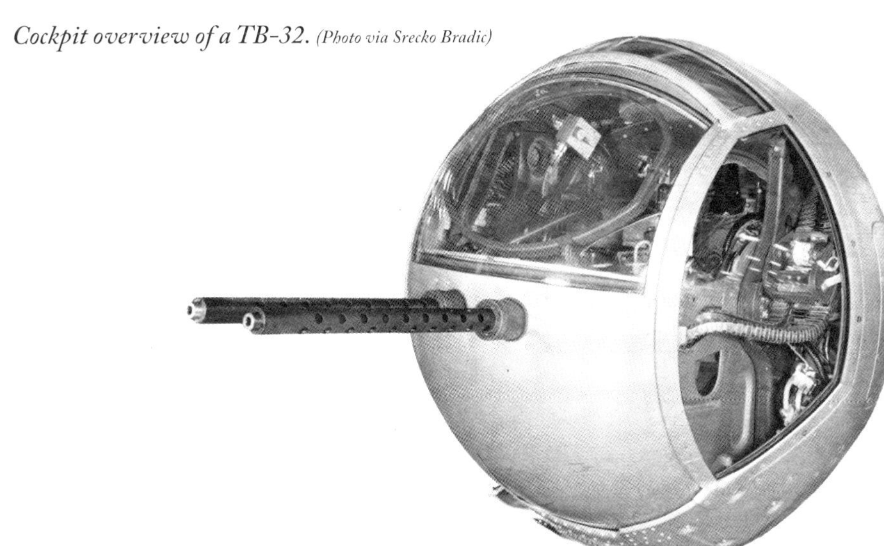

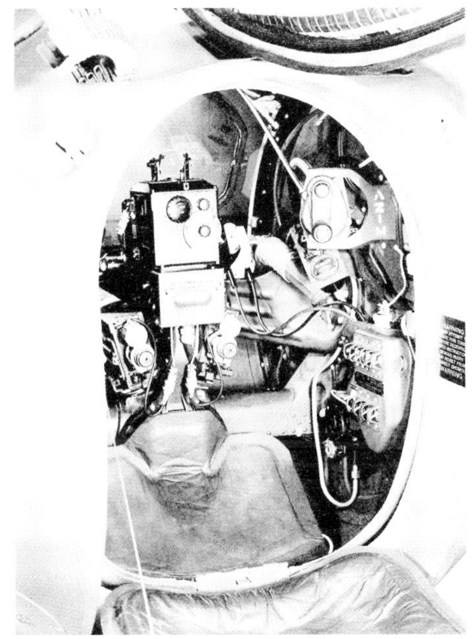

Exterior and interior of the Sperry A-17 nose turret. (Photos via Nick Veronico)

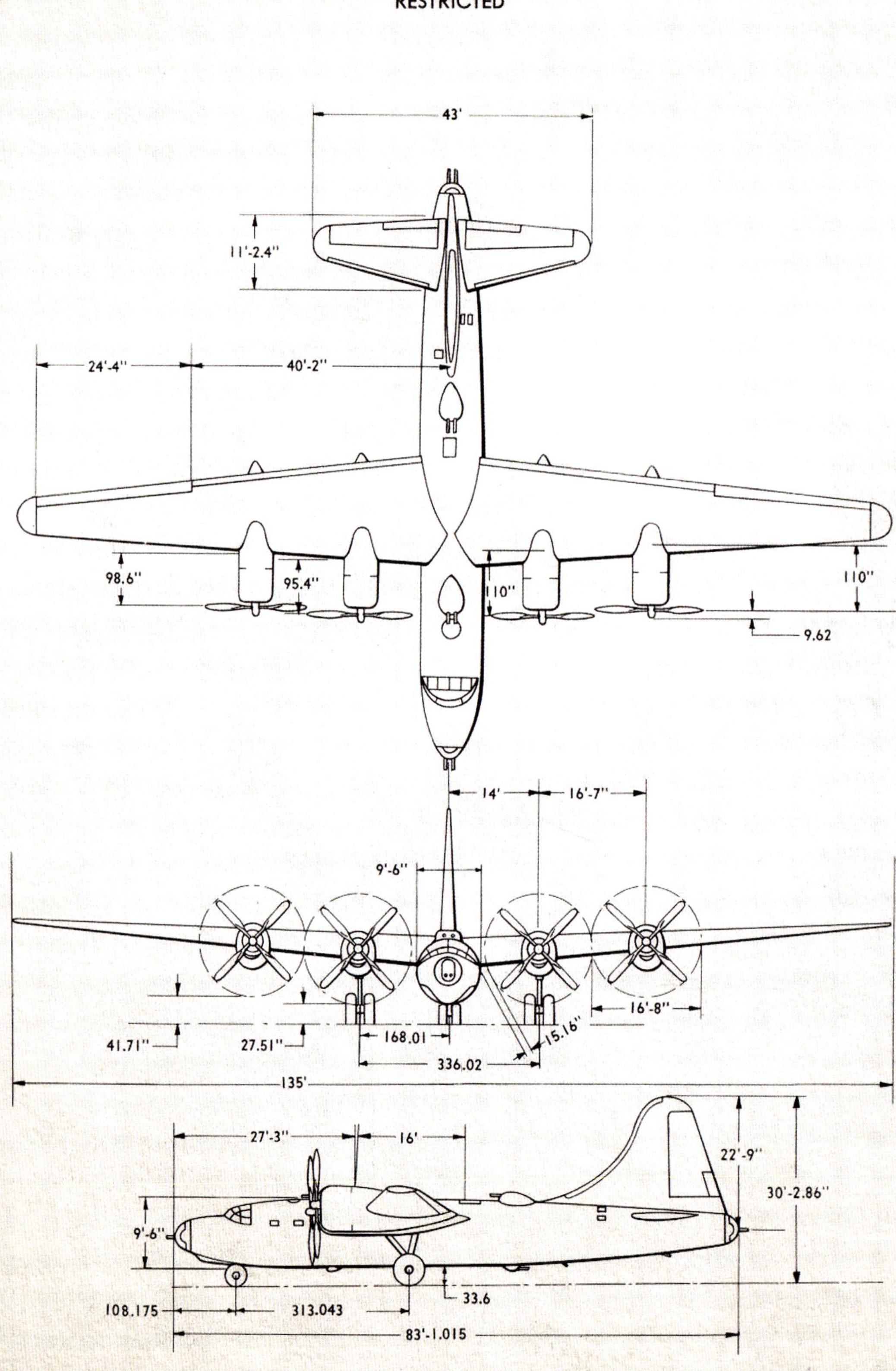

Figure 3 — Principal Dimensions

RESTRICTED

TECHNICAL DETAILS.

42-108476 during the early test phase of the B-32.

B-32 Dominator		
Wingspan	135 ft (41,15 m)	
Length	83 ft, 1 in (25,02 m)	
Height	32 ft (9,8 m)	
Wing surface	1,422 sq ft (132,1 m2)	
Empty weight	60,278 lbs (27.342 kg)	
Maximum take-off weight	100,800 lbs (45.722 kg)	
Top speed (30,000 ft)	357 mph (575 km/h)	
Top speed (5,000 ft)	302 mph (486 km/h)	
Cruise speed	290 mph (470 km/h)	
Rate of climb		
To 5,000 ft	850 ft/min	(6.0 min)
To 10,000 ft	750 ft/min	(12.0 min)
To 15,000 ft	650 ft/min	(19.0 min)
To 20,000 ft	500 ft/min	(27.5 min)
To 25,000 ft	400 ft/min	(38.0 min)

Service ceiling	30,700 ft (9,400 m)
Flight range	
Ferry range, no bomb load	4400 miles
10,000-pound bomb load	3000 miles
Bomb load options	
Single Suspension	40 100-pound M30 General Purpose 40 500-pound M64 General Purpose 12 1000-pound M65 General Purpose 8 1600-pound Mk 1 Armour Piercing 8 2000-pound M66 General Purpose 4 4000-pound M56 Light Case 40 500-pound T4E4 Delayed Opening Frag Clusters 36 500-pound E46 or E48 Incendiary Clusters 40 E28 or E36 Incendiary Clusters
Multiple Suspension	112 100-pound AN-M30 General Purpose, triple suspension 112 100-pound M12 Incendiary Clusters, triple suspension 136 100-pound M47 A2 Incendiary Bombs 120 100-pound M47 A2 Incendiary Bombs, triple suspension 80 260-pound AN-81 Frags, double suspension

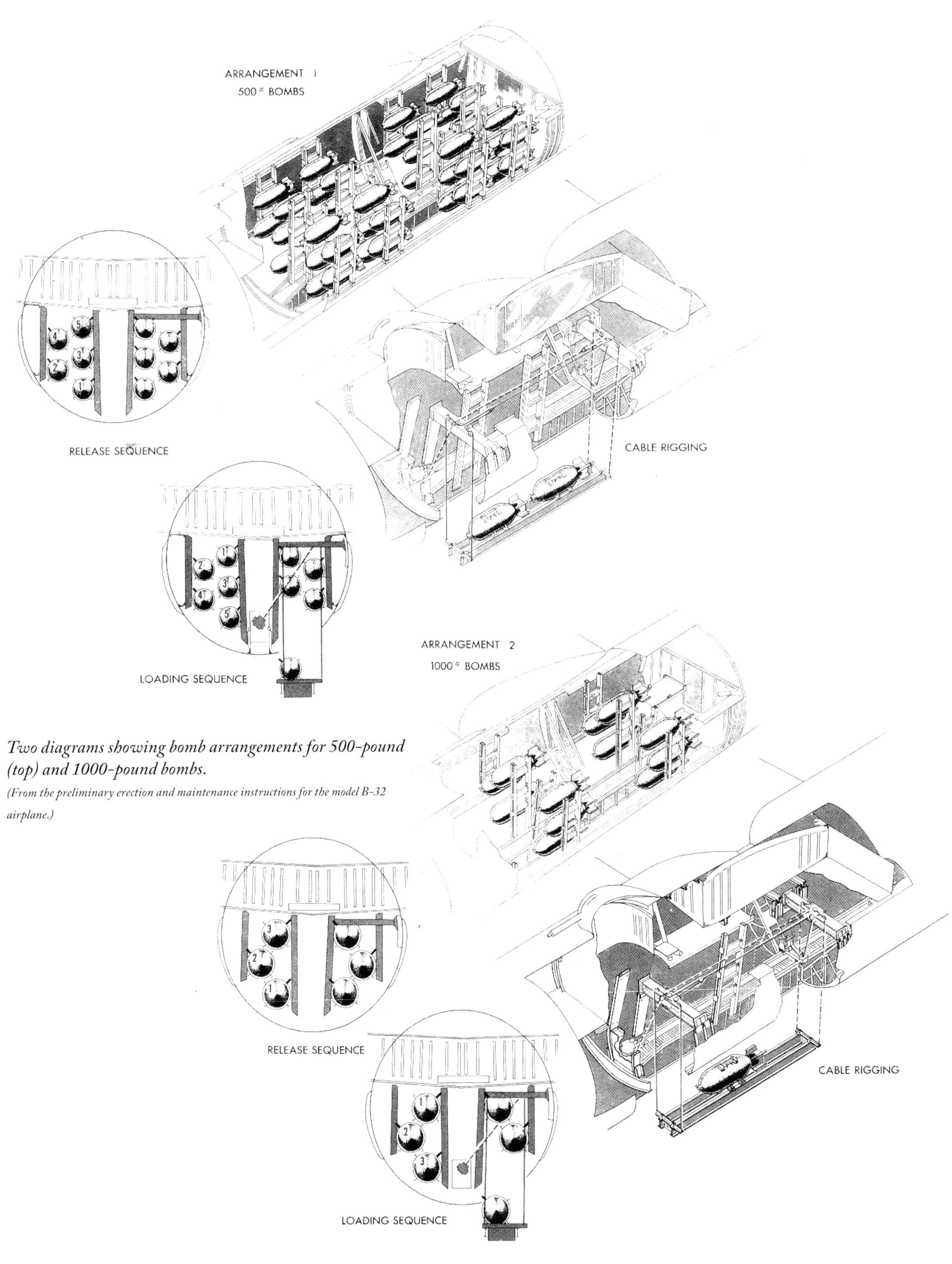

Two diagrams showing bomb arrangements for 500-pound (top) and 1000-pound bombs.

(From the preliminary erection and maintenance instructions for the model B-32 airplane.)

RESTRICTED

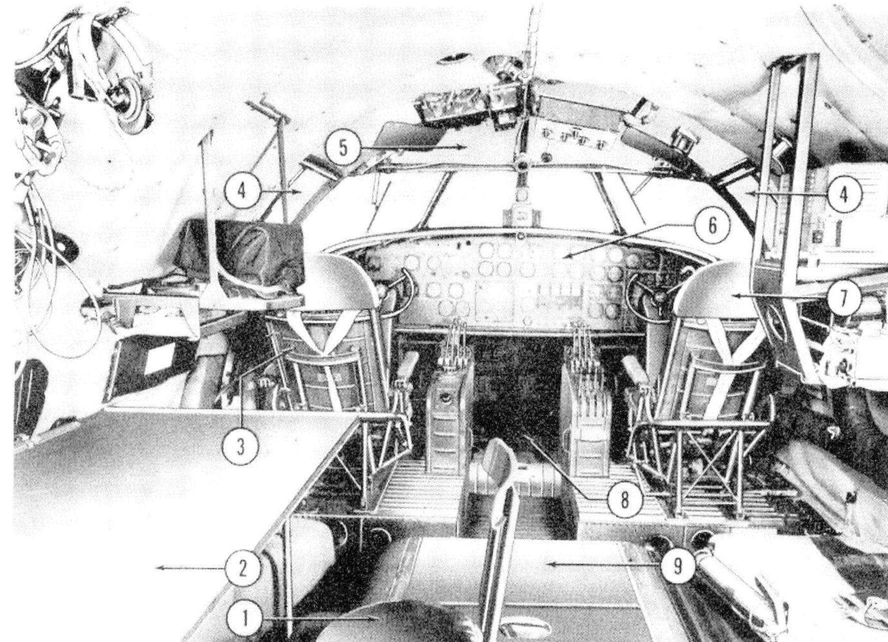

LEGEND

1. Navigator's Chair
2. Navigator's Table
3. Pilot's Chair
4. Sliding Enclosure Windows
5. Pilot's Enclosure
6. Main Instrument Panel
7. Copilot's Chair
8. Nose Compartment Passageway
9. Flight Deck Entrance

Figure 61 — Forward View of Flight Deck

LEGEND

1. Settee
2. Emergency Hydraulic Reservoir
3. Voltmeter Panel
4. Astroglass
5. Upper Forward Turret
6. Radio Operator's Table
7. Navigator's Panel
8. Bomb Bay Passageway
9. Navigator's Table

Figure 62 — Aft View of Flight Deck

RESTRICTED

Dedicated to Nico Braas (21 May 1946 – 27 June 2022)

Work on this book was started by our dear friend Nico Braas. Nico was instrumental in the early success of the Warplane series. His broad knowledge of aviation was truly inspiring. His passion for aviation endures in the numerous book titles he has written. We have taken great care to finish this book in the spirit of Nico's love for aviation. We have done our best to verify the sources of the photographs he selected. If you have any remarks on ownership, please do get in touch.

Edwin Hoogschagen – on behalf of the Walburg Lanasta team

Author
Edwin Hoogschagen

Publisher
Walburg Pers / Lanasta

Graphic design
Jantinus Mulder

Translation Revision
Deben Translations

First print, June 2025
ISBN 978-94-6456-571-3

NUR 465

Contact Warplane:
jantinusmulder@walburgpers.nl

Lanasta

www.walburgpers.nl/lanasta

While TB-32 42-108524 banks away from the camera ship it shows the VHF and IFF antennas and radio compass receiver fitted on the central fuselage beam.

Acknowledgement and thanks

This book would not have been possible without the generous support of Wolter Bonkestooter, Srecko Bradic, Pieto van Buysen, and Thijs Postma. A special word of thanks to Nick Veronico, who helped by providing unique photo material.

Pictures: from the author's collection, unless stated otherwise.

STUDIES, UNBUILT PROJECTS

Several B-32 variants were developed, among these were a commercial variant carrying up to 78 passengers (1941), a Troop transport for up to 130 troops or 92 paratroopers (1943), a cargo transport, capable of carrying 10,000 pounds of cargo (1943), a US Navy patrol bomber (1945). A turbo-propped variant was proposed in October 1944, which was to be powered with four General Electric TG-100 turboprop engines. Development was cancelled due to the limited power output of the engines.

Sources

- Butler, P., Hagedorn, D., Air Arsenal North America, Midland Publishing, 2004
- Harding, S., Long, J.L., Dominator. The Story of the B-32 Bomber, Pictorial Histories Publishing Co. U.S.A., 1986
- Harding, S., Flying Terminated Inventory, Wings, Vol. 23 No. 2, April 1993
- Head, W., Tindle, J., The B-32 Dominator Bomber and its Tragic History, Air Power History, Vol. 67 No. 1, Spring 2020
- Johnson, F.A., Last and Unluckiest of the Hemisphere Bombers, Wings, Vol. 4 No. 1, Feb 1974
- Mizrahi, J.V. (Editor), Requiem for Heavyweights, Wings, Vol 20 No.3, June 1990
- O'Leary, M., Consolidated B-24 Liberator – Production Line to Frontline, Osprey Publishing, 2002
- Sinko, B.A., Echoes of the Dominator, 2007
- Wolf, W., Consolidated B-32 Dominator, Schiffer Military History U.S.A., 2006
- Blood, W.T., Y., The Second String, A.A.H.S. Journal, 1968

- Consolidated Models 33/34: Genesis/Derivatives of the "Dominator" | Secret Projects Forum (secretprojects.co.uk)
- Dominator B-32A, Heavy Bomber (airpages.ru)
- Consolidated Aircraft (pilotfriend.com)
- My Private Story of the Davis Wing (earlyflightera.com)